International Environmental Labelling

Vol.1

For All Food Industries

(Meat, Beverage, Dairy, Bakeries, Tortilla, Grain and Oilseed, Fruit and Vegetable, Seafood, And Sugar and Confectionery)

Jahangir Asadi

Vancouver, BC CANADA

Copyright © 2020 by Top Ten Award International Network.

All rights reserved. No part of this publication may be reproduced, distributed or transmitted in any form or by any means, including photocopying, recording, or other electronic or mechanical methods, without the prior written permission of the publisher, except in the case of brief quotations embodied in critical reviews and certain other noncommercial uses permitted by copyright law. For permission requests, write to the publisher, addressed "Attention: Permissions Coordinator," at the address below.

Published by: Top Ten Award International Network
Vancouver, BC **CANADA**
Email: Info@TopTenAward.net
www.TopTenAward.net

Ordering Information:
Quantity sales. Special discounts are available on quantity purchases by universities, schools, corporations, associations, and others. For details, contact the "Sales Department" at the above mentioned email address.

International Environmental Labelling Vol.1/J.Asadi—1st ed.
ISBN 978-1-7773356-0-1

Contents

About TTAIN 10
Interoduction 13
General principles of environmental labelling 20
Types of environmental labelling 28
Types I environmental labelling 11
Types II environmental labelling 46
Types III environmental labelling 52
The meaning of Recycling 56
TTAIN environmental pioneers 74
International Organizations 80
Bibliography 85
Search by logos 92
Recycling Codes 100
Environmental friendly photos 102

I dedicate this book to my dear wife, with love and gratitude

We hope that, 10,000 years from now, future generations will be able to see flowers that provide bees with nectar and pollen and...
BEES provide flowers with the means to reproduce by spreading pollen from flower to flower,....

                                                                    J.Asadi

Acknowledgements:

I wish to thank my committee members, who were more than generous with their expertise and precious time. I would like to acknowledge and thank the Top Ten Award International Network for allowing me to conduct my research and providing any assistance requested. Special thanks goes to the members of the staff development and human resources department of Bia2Bc for their continued support.

It should be noted that all the required permissions for using the logos and trade marks has been obtained to be published in this volume.

The more you care about our environment, the more it will be protected from contaminants and toxins

About TTAIN

Top Ten Award International Network

Top Ten Award international Network (TTAIN) was established in 2012 to recognize outstanding individuals, groups, companies, organizations representing the best in the public works profession.
TTAIN publishing books related to international Eco-labeling plans to increase public knowledge in purchasing based on the environmental impacts of products.
Top Ten Award International Network provides A to Z book publishing services and distribution to over 39,000 booksellers worldwide, including Apple, Amazon, Barnes & Noble, Indigo, Google Play Books, and many more.
Our services including: editing, design, distribution, marketing
TTAIN Book publishing are in the following categories:

Student
Standard
Business
Professional
Honorary

We focus on quality, environmental & food safety management systems , as well as environmnetal sustain for future kids. TTAIN also provide complete consulting services for QMS, EMS, FSMS, HACCP and Ecolabeling based on international standards.

ISO 14024 establishes the principles and procedures for developing Type I environmental labelling programmes, including the selection of product categories, product environmental criteria and product function characteristics, and for assessing and demonstrating compliance. ISO 14024 also establishes the certification procedures for awarding the label.

TTAIN has enough experiences to help create new ecolabeling programmes in different countries all over the world.
For more detail visit our website : http://toptenaward.net
and/or send your enquiery to the following email:
info@toptenaward.net

CHAPTER 1

Introduction

This book is dedicated to the subject of environmental labels. The basis for the classification of its parts goes back to the types of environmental labelling according to the classifications provided by the International Organization for Standardization. In each section, while presenting the relevant definitions, I mention the existing international standards and present examples related to each type of labelling. Environmental labelling is an important and significant topic, and its richness is added to every day, which has attracted the attention of many experts and researchers around the world. The idea of compiling this book, came to my mind when I observed that national environmental labelling models have been developed in most countries of the world, but in many other countries, the initial steps have not been taken yet. Therefore, I decided to create the first spark for the development of environmental labelling patterns in other countries by collecting appropriate materials and inserting samples of labelling patterns of different countries of the world. It should be noted that the description of each environmental label in this book does not indicate their approval or denial; they are included only to increase the awareness of all enthusiasts and consumers of the meanings and concepts derived from such labels. We hereby ask all interested parties around the world who wish to start an environmental labelling program in their country to benefit from our intellectual assistance and support in the form of consulting contracts. Increasing human awareness of the urgent need to protect the environment has led to changes in all levels of activities, including the production of marketing products, consumption, use, and sale of goods and services at the national and international levels. Stakeholders involved in environmental protection include consumers, producers, traders, scientific and technological institutes, national authorities, local and international organizations, environmental gatherings, and human society in general. Decisions by consumers and sellers of products are made not only on the basis of key points such as quality, price, and availability of

products but also on the environmental consequences of products, including the consequences that a product can have before, after and during production. The most important environmental consequences include water, soil, and air pollution along with waste generation, especially hazardous waste. Further consequences include noise, odor, dust, vibration, and heat dissipation as well as energy consumption using water, land, fuel, wood, and other natural resources. There are further effects on certain parts of the ecosystem and the environment. In addition, the environmental consequences not only include the natural use of the products but also abnormal and even emergency or accidental uses. The basis of studies and studies in this field is done through product life cycle evaluation, which generally involves the study and evaluation of environmental aspects and consequences of a category (product, service, etc.) because of the preparation of raw materials for production until they are used or discarded. Sometimes the phrase "review from cradle to grave" is used for such an evaluation. In addition to the above, the environmental consequences that may occur at any stage of the product life cycle, including the preliminary stages and its preparation, production, distribution, operation, and sale, should also be considered when evaluating it. This type of evaluation refers to product life cycle analysis from an environmental point of view,"

which is a useful tool for measuring the degree of environmental health of a product, comparing different products, improving product quality, and confirming the environmental health claims of the product. The environmental health analysis tool for products and services facilitates their placement in domestic or foreign markets, considering that the awareness of consumers and retailers about the environmental consequences of the product has increased, as has the accurate and explicit measurement by the people in charge at all levels. Local, national, and international in the field of environmental protection. Products that can claim to be environmentally complete in all stages of their life cycle and meet the mandatory and optional environmental needs are considered successful products. Environmental messages refer to the policies, goals, and skills of product manufacturing companies as part of the environmental management systems in which they are applied, and consumers and retailers are increasingly paying attention to this issue when making purchasing decisions. In addition, companies have been encouraged and even forced to adapt their environmental management systems to agencies and retailers and to local, national, international, and other environmental issues.

The environmental health message of a product can be conveyed to the consumer in various ways, including implicitly or explicitly. For example, the implicit or implicit message conveyed directly by the product to the customer is that the product is suitable for the intended use and purpose, and, without material waste in size, weight, and dimensions, is perfectly proportioned and without additional packaging. Sometimes it is necessary to convey these messages and claims about the correctness of the product quite clearly through magazines or other media as well as through certificates that are accurate, simple, and convincing to the consumer in the form of a label. These messages must be accurate and fact-based; otherwise they will nullify the product and create contradictory effects. Confirmation of these claims by a third-party organization will increase its credibility. It should also be noted that the multiplicity of these messages, depending on the type of products or companies producing them, confuses consumers in the market and also creates artificial boundaries or causes a differentiated distinction against certain products or companies. Various models, principles, and methods have been provided by local, regional, national, and international organizations to demonstrate product life cycle analysis and other guidelines on environmental management systems and their labels. At the national level, significant advances have been made in the design of environmental labels in various countries, including developing countries and the Scandinavian countries. For example, the first project was designated in Germany as a Blue Angel in 1977, later on Canada in 1988, the Scandinavian countries and Japan in 1989, the United States and New Zealand in 1990, India, Austria, and Australia in 1991, And in 1992, Singapore, the Republic of Korea, and the Netherlands developed their national environmental labelling. Environmental labels are an environmental management tool that is the subject of a series of ISO 14000 standards. These environmental labels provide information about a product or commodity in terms of its broad environmental characteristics, whether it is about a specific environmental issue or about other characteristics and topics.Interested and pro-environmental buyers can use this information when choosing products or goods. Product makers with these environmental labels hope to influence people's purchasing decisions. If these environmental labels have this effect, the share of the product in question can increase, and other suppliers may create healthy environmental competition by improving the environmental aspects of their products and commodities. The overall goal of environmental labels is to convey acceptable and accurate information that is in no way misleading regarding the environmental aspects of products and commodities, and they encourage the consumer to buy and produce products that reduce stress on the environment. Environmental labelling must follow the general principles that the International Organization for Standardization has published in a collection entitled the ISO 14020 standard, which refers to these general principles here. It should be noted that other documents and laws in this field are considered if they are in accordance with the principles set out in ISO 14020.

Along with QUALITY and PRICE, consumers consider products' Environmental impact

CHAPTER 2

General Principles on Environmental Labelling

1 The First Principle: Evironmental notices and labels must be accurate, verifiable, relevant, and in no way misleading and/or deceptive.

2 The Second Principle: Procedures and requirements for environmental labels will not be ready for selection unless they are implemented by affecting or eliminating unnecessary barriers to international trade.

3 The Third Principle: Environmental notices and labels will be based on scientific analysis that is sufficiently broad and comprehensive, and to support this claim, the product must be reliable and reproducible.

4 The Fourth Principle: The process, methodology, and any criteria required to support the announcements on environmental labels will be available upon request all interested groups.

5 The Fifth Principle: Development and improvement of environmental notices and labels should be considered in all aspects related to the service life of the product.

6 The Sixth Principle: Announcements on environmental labels will not prevent initiative and innovation but will be important in maintaining environmental implementation.

7 The Seventh Principle: Any enforcement request or information requirement related to environmental notices and labels should be limited to the necessary information to establish compliance with an acceptable standard and based on the notification standards and environmental labels.

8 The Eighth Principle: The process of improving the announcement and environmental labels should be done by an open solution with interested groups. Reasonable impressions must be made to reach a consensus through this process.

9 The Ninth Principle: Information on the environmental aspects of the product and goods related to an advertisement and environmental label will be prepared for buyers and interested buyers from a group consisting of an advertisement and an environmental label.

Activities that will teach kids about sustainability: Recycle, Pick up trash, Sort the garbage, Finding Ecolabels on products during shopping...

CHAPTER 3

Types of Environmental Labelling

At present, according to the classification provided by the International Organization for Standardization, there are three types of environmental labelling patterns:

1. Type I labelling: This labelling is known as eco-labelling, and because it is difficult to translate this word into many languages, it presents another reason to adhere to a numerical classification system. In the content of Type I labelling, a set of social commitments that creates criteria according to the scientific principles on the basis of which a product is environmentally preferable is discussed. Consumers are then instructed in assessing environmental claims and must decide which packaging is more important.

2. Type II labelling: refers to the claims made on product labels in connection with business centers. This includes familiar claims such as recyclable, ozone-friendly, 60% phosphate-free, and the like. This type of labelling can be in the form of a mark or sentence on the product packaging. Some of them are valid environmental claims—and some can be completely misleading. Usually, all countries have laws against deceptive advertisements, so why has the International Organization for Standardization discussed this issue? The answer is that it is not clear whether the environmental claims have a technical basis or whether the ad is meaningless.

3. Type III labelling: is a distinct form of third-party environmental labelling pattern designed to avoid the difficulties that can result from type-one labelling. Technical committee for Environment of International organization for Standardization has undertaken a new project to standardize guidelines and Type III labelling methods. One of the main objections raised by industries to Type I labelling is the basis for its management

Farm to School, is a school based program linking school children from Kindergarten to grade 12 with local farms.

CHAPTER 4

Type I Environmental Labelling

Type I labelling: This labelling is known as eco-labelling, and because it is difficult to translate this word into many languages, it presents another reason to adhere to a numerical classification system. In the content of Type I labelling, a set of social commitments that creates criteria according to the scientific principles on the basis of which a product is environmentally preferable is discussed. Consumers are then instructed in assessing environmental claims and must decide which packaging is more important.

Type I adhesive has the following specifications:
A. Has an optional third-party template.
B. When the product meets a certain standard, the labelling of this product is included.
C. The purpose of this program is to identify and promote products that play a pioneering role in terms of environment, which means its criteria are at a higher level than the average environmental performance.
D. Acceptance/rejection criteria are determined for each group of products and are publicly available.
E. The criteria are adjusted after considering the environmental consequences of the product life cycle.

Examples of Type I Labelling:
In this section, and considering the importance of this type of labelling, I provide a description of some examples of Type I labelling related to some countries along with a list of products on which this mark is placed.

New Zealand

Environmental Choice New Zealand (ECNZ) is the country›s only Government-owned ecolabel. Administered by the New Zealand Ecolabelling Trust, the ecolabel was established in 1992 to provide a credible and independent guide for businesses and consumers to purchase and use products that are better for the environment.

A member of the Global Ecolabelling Network, ECNZ is a Type I ecolabel, which means products and services bearing the label meet criteria covering the whole life-cycle of the product/service, from raw materials, through manufacture and usage, to end-of-life disposal or reuse. Licensed products and services are independently assessed regularly by a third party.

The New Zealand Ecolabelling Trust
PO Box 56 533, Dominion Rd, Mt Eden, Auckland 1446
Tel: 0064 9 845 3330
Email: info@environmentalchoice.org.nz
Web: www.environmentalchoice.org.nz

Singapore

Established in 1995, the Singapore Environment Council (SEC) is an independently managed, non-profit and non-governmental organisation (NGO). As Singapore's first United Nations Environment Programme (UNEP)-accredited environmental NGO, we influence thinking of sustainability issues and coordinate environmental efforts in the nation.

We are also an approved charity and offer tax exemption to donors. SEC continuously engages all sectors of the community by formulating and executing a range of holistic programmes, such as the Singapore Environmental Achievement Awards, Asian Environmental Journalism Awards, School Green Awards, Singapore Green Labelling Scheme, Project: Eco-Office, Project: Eco-Shop and Project: Eco-F&B. In addition, we build a pool of committed volunteers under our Earth Helpers programme. Our Training and Education arm provides the people, public and private sectors with the opportunity to develop awareness, knowledge, skills and tools in order to protect and improve our environment for a sustainable future.

Strong partnerships with corporations, government agencies and other NGOs are valued by us. These partnerships are vital for sustaining our programmes, leading to positive action and change. Over the years, SEC has given strength and direction to the environmental movement in Singapore.

For further information, please visit https://sec.org.sg/.

Philippines

The National Ecolabelling Programme Green Choice Philippines (NELP-GCP) is an ecolabelling programme based on ISO 14024 Guiding Principles and Procedures. It is a voluntary, multiple criteria-based, and third-party programme the aims to encourage clean manufacturing practices and consumption of environmentally preferable products and services. It awards the seal of approval to product or service that meets the environmental criteria established for the product category by a multi-sector Technical Committee. Products with the Green Choice Philippines Seal assures the consumers on its preference for the environment. NELP-GCP is being administered by the Philippine Center for Environmental Protection and Sustainable Development, Inc. (PCEPSDI).

Contact:
Website: https://pcepsdi.org.ph/
E-mail: greenchoicephilippines@pcepsdi.org.ph,
greenchoicephilippines@gmail.com

Denmark, Finland, Norway, Iceland, Sweden

The Nordic Swan Ecolabel
The Nordic Swan Ecolabel is the official Nordic ecolabel supported by all Nordic Governments. It is among the world's strictest and most recognised environmental certifications.

The Nordic Swan Ecolabel is a Type I environmental labelling program established in 1989 by the Nordic Council of Ministers, connect¬ing policy, people, and businesses with the mission to make it easy to make the environmentally best choice. Nordic Ecolabelling is the non-profit organisation responsible for the Nordic Swan Ecolabel.

The organisation offers independent third-party certification and support for a wide range of product areas and services, ensuring that they comply with the Nordic Swan Ecolabel's strict requirements through documentation and inspections.

30 years of experience and expertise has made the Nordic Swan Ecolabel a powerful tool that paves the way to a sustainable future by giving producers a recipe on how to develop more environmentally sustainable products, and giving consumers credible guidance by helping them identify products that are among the environmentally best.

Globally, you can find more than 25,000 Nordic Swan ecolabelled products. 93% of all Nordic consumers recognise the Nordic Swan Ecolabel as a brand, and 74% believe that the Nordic Swan Ecolabel makes it easier for them to make envi¬ronmentally friendly choices (IPSOS 2019).

Denmark, Finland, Norway, Iceland, Sweden

Securing a sustainable future
The Nordic Swan Ecolabel works to reduce the overall environmental impact from production and consumption and contributes significantly to UN Sustainable Development Goal 12: Responsible consumption and production.
To ensure maximum environmental impact, the Nordic Swan Ecolabel sets product specific requirements and evaluates the environmental impact of a product in all relevant stages of a product lifecycle - from raw materials, production, and use, to waste, re-use and recycling.
Common to all products certified with the Nordic Swan Ecolabel is that they meet strict environmental and health requirements. All requirements must be documented and are verified by Nordic Ecolabelling. Nordic Ecolabelling regularly reviews and tightens the requirements.
Therefore, certifications are time-limited and companies must re-apply to ensure sustainable development.

International website:
Nordic-ecolabel.org
National websites:
Denmark: ecolabel.dk
Sweden: svanen.se
Norway: svanemerket.no   (in Norwegian)
Finland: joutsenmerkki.fi (in Finnish)
Iceland: svanurinn.is   (in Icelandic)

Canada

The Marine Stewardship Council (MSC) is a global nonprofit organization established to protect the last major food resource that is truly wild: seafood.

The MSC works with fisheries, grocery stores, restaurants, and other companies to change the way the oceans are fished, address food fraud, and recognize and reward sustainable fishing practices.

The MSC blue fish ecolabel makes it simple for consumers to identify seafood options that are wild, certified sustainable, and traceable from ocean to plate.

To learn more, visit msc.org.

Contact:
Website: https://msc.org

Thailand

The Thai Green Label Scheme was initiated by the Thailand Business Council for Sustainable Development (TBCSD) in October 1993. It was formally launched in August 1994 by The Thailand Environment Institute (TEI) and Thai Industrial Standards Institute (TISI). The Green Label is an environmental certification logo awarded to specific products which have less detrimental impact on the environment in comparison with other products serving the same function. The Thai Green Label Scheme applies to all products and services, but not foods, beverage, and pharmaceuticals. Products or services which meet the Thai Green Label criteria may carry the Thai Green Label. Participation in the scheme is voluntary.

Thailand Environment Institute (TEI)
16/151 Muang Thong Thani, Bond Street,
Bangpood, Pakkred, Nonthaburi 11120 THAILAND
Tel. +66 2 503 3333 ext. 303, 315, 116
Fax. +66 2 504 4826-8
Website: http://www.tei.or.th/greenlabel/
Email: lunchakorn@tei.or.th

USA

The International Marine Mammal Project (IMMP) of Earth Island Institute works to protect whales and dolphins and their ocean habitats. In 1990, IMMP established the strong Dolphin Safe tuna standards and logo to guide consumers to buying tuna that was not caught by chasing, netting, and killing dolphins. More than 90% of the global tuna industry is part of our program. IMMP also has monitors around the world who check tuna supplies to verify tuna is caught without harm to dolphins and other marine life. Our program prevents the deaths of more than 90,000 dolphins annually.

Earth Island Institute
2150 Allston Way, Suite 460
Berkeley, California USA
(510) 859-9100
Web: www.savedolphins.eii.org

Peru

BIO LATINA, the consolidated byproduct of four Latin American national certification entities. Since 1998, we have provided certification services in Latin America for national and international markets. We seek to help create a more sustainable and resilient world. With these goals in mind, we have expanded our service portfolio beyond organic to social and environmental certifications.

Visit us: https://biolatina.com

From our regional offices we serve Latin American.

Our headquaters:
Av. Javier Prado Oeste 2501, Bloom Tower Of. 802, Magdalena del Mar, Lima 17, Perú

ORGANIC CERTIFICATION

Lithuania

EKOAGROS is the only institution in Lithuania for more than 20 years carrying out certification and control activities of organic production and products of national quality, also providing services of certification activities in accordance with the foreign national and private standards in foreign countries. From year 2017 EKOAGROS is accredited as certifying agent to conduct certification activities on crops, wild crops, livestock and handling operations in accordance with USDA NOP.

Contact information:
EKOAGROS
Address K. Donelaicio str. 33, LT-44240 Kaunas, Lithuania
Tel. No. +370 37 20 31 81
Website: www.ekoagros.lt

Germany

FSC® is a global not-for-profit organization that sets the standards for responsibly managed forests, both environmentally and socially. When timber leaves an FSC certified forest they ensure companies along the supply chain meet our best practice standards also, so that when a product bears the FSC logo, you can be sure it's been made from responsible sources. In this way, FSC certification helps forests remain thriving environments for generations to come, by helping you make ethical and responsible choices at your local supermarket, bookstore, furniture retailer, and beyond. www.fsc.org

FSC® International
Adenauerallee 134
53113 Bonn
E-mail: info@fsc.org
Phone: +49 (0) 228 367 66

FSC Canada
50 rue Sainte-Catherine Ouest,
bureau 380B, Montreal, QC H2X 3V4
Email: info@ca.fsc.org
Telephone: 514-394-1137

USA

The Carbonfree® Product Certification is a meaningful, transparent way for you to provide environmentally-responsible, carbon neutral products to your customers. By determining a product's carbon footprint, reducing it where possible and offsetting remaining emissions through our third-party validated carbon reduction projects, companies can:
- Differentiate their brand and product
- Increase sales and market share
- Improve customer loyalty
- Strengthen corporate social responsibility & environmental goals

The Carbonfree® Product Certification Program is proud to be part of Amazon's Climate Pledge Friendly Program!
Carbonfund.org is leading the fight against climate change, making it easy and affordable to reduce & offset climate impact and hasten the transition to a clean energy future.

Contact:

O: 240.247.0630 ext 633
C: 203.257.7808
M: 853 Main Street, East Aurora, NY, 14052

EUROPE

Established in 1992 and recognized across Europe and worldwide, the EU Ecolabel is a label of environmental excellence that is awarded to products and services meeting high environmental standards throughout their life-cycle: from raw material extraction, to production, distribution and disposal. The EU Ecolabel promotes the circular economy by encouraging producers to generate less waste and CO2 during the manufacturing process. The EU Ecolabel criteria also encourages companies to develop products that are durable, easy to repair and recycle.

The EU Ecolabel criteria provide exigent guidelines for companies looking to lower their environmental impact and guarantee the efficiency of their environmental actions through third party controls. Furthermore, many companies turn to the EU Ecolabel criteria for guidance on eco-friendly best practices when developing their product lines. The EU Ecolabel helps you identify products and services that have a reduced environmental impact throughout their life cycle, from the extraction of raw material through to production, use and disposal. Recognised throughout Europe, EU Ecolabel is a voluntary label promoting environmental excellence which can be trusted.

Spain , Germany, Italy, Sweden, Greece, Portugal, Poland, Belgium, Netherlands, Estonia, Finland, Austria, Lithuania, Czech Republic, Norway, Cyprus, Ireland, Slovenia, Hungary, Romania, Croatia, Bulgaria, Malta, Slovak Republic, Latvia, Luxembourg, Iceland

Contact and more information via: http://ec.europe.eu

Sweden

TCO Certified is the world-leading sustainability certification for IT products. Covering 11 product categories including computers, mobile devices, display products, and data center products, its comprehensive criteria are designed to drive social and environmental responsibility throughout the product life cycle. Independent verification of criteria compliance is always included. Independent verifiers spend around 20,000 hours every year on tests and audits. Currently, more than 3,500 products from 27 well-known IT brands are certified. The purpose of TCO Certified is to drive progress toward a future where all IT products have a sustainable life cycle, something that requires a collective effort from IT buyers as well as industry. TCO Certified helps the IT industry structure their work with sustainability and offers a platform for continuous improvement. Organizations that buy IT products use the certification as a tool for making more responsible IT product choices.

Contact:
TCO Development |
Linnégatan 14 | 11447 Stockholm, Sweden |
Mobile: +46 (0) 706 358351|
Email: Marketing@tcodevelopment.com

Netherland

For more than 25 years, the independent Dutch foundation SMK works from professional knowledge with companies to improve the sustainability of products and business management. SMK cooperates with an extensive stakeholder network of governments, producers, branch and non-governmental organisations, retailers, consultancies, researchers. The SMK Boards of Experts establish objective criteria for more sustainable products and services. SMK's transparent work processes, third party audits and certifications are conducted according to international certification standards, mostly under supervision of the Dutch Accreditation Council. Besides, SMK is Competent Body of the EU Ecolabel. SMK keeps an extensive database of sustainability criteria.

Contact:
Bezuidenhoutseweg 105 - 2594 AC Den Haag
Telefoon: 070-3586300
Mobiel: 06-82311031
(niet op woensdag)
www.smk.nl

CHAPTER 5

Type II Environmental Labelling

Type II environmental labelling refers to the claims made on product labels in connection with business centers. This includes familiar claims such as recyclable, ozone-free, 60% phosphate-free, and the like. This type of labelling can be in the form of a mark or sentence on the product packaging. Some of them are valid environmental claims—and some can be completely misleading.

Usually, all countries have laws against deceptive advertisements, so why has the International Organization for Standardization discussed this issue? The answer is that it is not clear whether the environmental claims have a technical basis or whether the ad is meaningless.

Most countries have guidelines at the national level to help producers and consumers know what constitutes a true, scientifically valid claim.
There is a national standard on this in Canada. In Australia, the Consumer Commission has published guidance on this, and there are similar examples in other countries.

USA

The original recycling symbol was designed in 1970 by Gary Anderson, a senior at the University of Southern California as a submission to the International Design Conference as part of a nationwide contest for high school and college students sponsored by the Container Corporation of America. The recycling symbol is in the public domain, and is not a trademark. The Container Corporation of America originally applied for a trademark on the design, but the application was challenged, and the corporation decided to abandon the claim. As such, anyone may use or modify the recycling symbol, royalty-free.

For More information refer to ISO 14021,
Environmental Labels and declarations

Canada

Environmental Sustain for Furure kids established in Vancouver, BC Canada in 2020. (ESFK) is an international ecolabel focused on taking care of environment for future of kids.

ESFK defined as 'self-declared' environmental claims made by manufacturers and businesses based on ISO 14020 series of standards, the claimant can declare the environmental objectives and targets in relation to taking care of environment for future kids. However, this declaration will be verifiable.

Environmental Sustain for Future Kids
Vancouver, BC CANADA

Email: info@esfk.org
Web: www.esfk.org

CHAPTER 6

Type III Environmental Labelling

Type III environmental labelling is a distinct form of third-party environmental labelling pattern designed to avoid the difficulties that can result from type-one labelling. Technical committee for Environment of International organization for Standardization has undertaken a new project to standardize guidelines and Type III labelling methods. One of the main objections raised by industries to Type I labelling is the basis for its management.

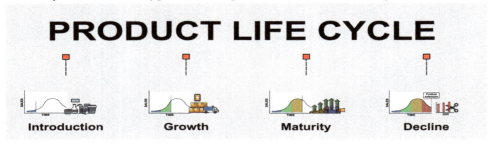

Due to the nature of the system, less than 50% of the various products on the market can meet the criteria and qualify for Type I Labelling. As long as the industry is the main supporter of other third-party models for quality systems, it is sometimes difficult for an industry to support a program that can only benefit 15% of its members. This type of labelling is currently practiced in some countries, such as Sweden, Canada, and the United States. Choosing the right product has never been easy, but Type III labelling will help because each product can have a label that describes its environmental performance and is certified by a third-party company. Consumers can then compare labels and choose their favorite products.

Life cycle assessment of supermarket food waste

The life cycle assessment results reported that the annual wastage of bread and beef products have the largest contribution to the environmental footprint of the supermarket.

Food supply chains are one of the main contributors to several pressing environmental problems such as climate change, eutrophication and loss of biodiversity.

All food follows a pathway through time and space – or a life cycle – from agricultural production through to eventual consumption or waste disposal. At each stage in the life cycle, resources are used (e.g. land) and outputs are created, some desirable (e.g. food), others not so (e.g. greenhouse gases).

CHAPTER 7

The Meaning of Recycling

As an average person drinks 60,000 liters in a lifetime, we are recycling water ourselves. While polluting our oceans we are poisoning our children and grandchildren. Water is natures most vital resource. Let's work together to prevent further pollution of our oceans. Currently, just 25% of mobile phones can be recycled right now. A total of 3.4 million tonnes of plastics were consumed in Australia. A total of 320 000 tonnes of plastics were

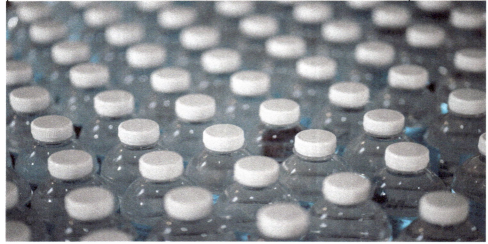

recycled, which is an increase of 10 per cent from the 2016-17 recovery. Recycling just one ton of aluminum cans conserves more than 152 million Btu, the equivalent of 1,024 gallons of gasoline or 21 barrels of oil consumed. The universal recycling symbol, logo or icon is an internationally recognized symbol used to designate recyclable materials. The recycling symbol is in the public domain and is not a trademark.

Step Up: Recycle, Reuse & Reduce
How can I recycle?

STEP 1: Go Green & Recycle
Separate recyclables from your trash with a recycling bin. Such as a dual trash can with two compartments or a compost bin. But you can also use a paper shredder for sorting sensitive paperwork that needs shredding before you recycle it. By using these recycling tools you prevent the loss of recyclable materials.

STEP 2: Go Green & Reuse
Plastic disposables and single-use products are wasteful and not stylish. Be fashionable and more eco-friendly and bring your own beautiful and reusable essentials such as a reusable water bottle, coffee cup or foldable shopping bag. Make a trendy statement and stop plastic pollution.

STEP 3: Go Green & Reduce
How do you reduce plastics and prevent plastic pollution? The answer is: use more natural resources. This is also known as living a zero waste lifestyle where the mission is to reduce waste as much as possible. Check our zero waste store with eco-friendly paper straws, bamboo toothbrushes, and metal safety razors.

STEP 4: Team Up & Go Green
Start a Green Team in your office or workplace together with your colleagues to educate, inspire, challenge and empower employees about your sustainability goals. Know what you throw away today in the office and think about how you and your colleagues can reduce, reuse and recycle tomorrow.

Why is a green lifestyle important?

Nature can't digest plastics because this material is not biodegradable. We can use much more natural resources that are biodegradable by nature itself. Because not 100% of what we consume will be collected or recycled.

The three arrows of the recycling symbol represent the three main stages of the recycling process: recycling, reusing and reducing. These three chasing arrows are also known as the recycling trilogy, and Together the arrows form a closed loop that symbolizes a circular economy.

What is recycling?

Recycling is the process of collecting and processing materials that would otherwise be thrown away as trash and turning them into new products. Recycling can benefit your community, the economy and the environment.

Is recycling truly beneficial for the environment?

The answer is Yes, For example:

Recycling 500 kg of paper can save the energy equivalent of consuming 160 gallons of gasoline.

Recycling just 1000 kg of aluminum cans conserves 1,000 gallons of gasoline or more than 20 barrels of oil consumed.

Plastic bottles are the most recycled plastic product in the United States as of 2015, according to our most recent report. Recycling just 10 plastic bottles saves enough energy to power a laptop for more than 25 hours.

Why is it important to only put items that can be recycled in the recycling bin?

Putting items in the recycling bin that can't be recycled can contaminate the recycling stream. After these unrecyclable items arrive at recycling centers, they can cause costly damage to the equipment. Additionally, after arriving at recycling centers, they must be sorted out and then sent to landfills, which raises costs for the facility. That is why it is important to check with your local recycling provider to ensure that they will accept certain items before placing them into a bin. Some items may also be accepted at retail locations or other at local recycling centers.

Why are some items that look recyclable not accepted at my recycling facility?

Your local recycling facility might not accept all recyclable items. This is especially true with plastics. While plastic bottles are the most commonly recycled plastic products, other plastics may or may not be accepted in your area, so first check what your local recycling provider accepts. It is important to understand that the existence of a plastic resin code on the product does not guarantee that the product is recyclable in your area. Additionally, glass may not be accepted in some areas, so please confirm with your local provider.

What should I never put in my recycling bin(s)?
Garden hoses
Sewing needles
Bowling balls
Food or food-soiled paper
Propane tanks or cylinders
Aerosol cans that aren't empty

Many communities have collection programs for household hazardous waste to reduce the potential harm posed by these chemicals.

What are the most common items that I can put into my curbside recycling bin?

Cardboard
Paper
Food boxes
Mail
Beverage cans
Food cans
Glass bottles
Jars (glass and plastic)
Jugs
Plastic bottles and caps

Generally, these are the most commonly recycled items. Please confirm with your local recycling provider first before putting these items in your curbside recycling bin, however, since what is accepted depends on your area.

Are paper or plastic shopping bags better for the environment? How about reusable bags versus disposal bags?

We do not have information on the environmental benefits of paper versus plastic bags. We encourages consumers to:

Reduce the number of bags they use,
Reduce the number of bags they throw away after one use,
Reuse bags, and
Recycle bags when they can no longer be used.
Consumers also can reduce waste by using reusable shopping bags.

Glass

This symbol asks that you recycle the glass container. Please dispose of glass bottles and jars in a bottle bank, remembering to separate colours, or use your glass household recycling collection if you have one.

WHY CAN'T I RECYCLE SOME GLASS ITEMS?
Some types of glass do not melt at the same temperature as bottles and jars. If they enter the glass recycling process it can result in new containers being rejected. These items should be recycled separately - check with your local household waste recycling centre.

HOW TO RECYCLE GLASS BOTTLES AND JARS
Put lids and caps back on. This reduces the chance of them getting lost during the sorting process as they can be recycled separately.
Empty and rinse - a quick rinse will do. Leftover liquid can contaminate other recyclables which may mean they aren't recycled.

Aluminium
This symbol indicates that the item is made from recyclable aluminium.

HOW TO RECYCLE
Rinse or wipe off any crumbs or food residue from foil trays. To rinse just dunk the tray in the washing up water - no need to run the tap.
Scrunch kitchen foil, tub and pot lids and wrappers together to form a ball - the bigger the ball, the easier it is to recycle.
As well as foil, you can usually recycle these other aluminium items:

Drinks cans
Screw top lids from bottles
(recycle with the bottle - the cap can be left on)
Takeaway containers and barbeque trays.

Symbol#1: PET or PETE

PET or PETE (polyethylene terephthalate) is the most common plastic for single-use bottled beverages, because it›s inexpensive, lightweight, and easy to recycle. It poses low risk of leaching breakdown products. Its recycling rates remain relatively low (around 20%), even though the material is in high demand by manufacturers.

Found in:
Soft drinks, water, ketchup mouthwash bottles; peanut butter containers; salad dressing and vegetable oil containers

How to recycle it:
PET or PETE can be picked up through most curbside recycling programs as long as it›s been emptied and rinsed of any food. When it comes to caps, our environmental pros say it's probably better to dispose of them in the trash (since they›re usually made of a different type of plastic), unless your town explicitly says you can throw them in the recycle bin. There›s no need to remove bottle labels because the recycling process separates them.

Recycled into:
Polar fleece, fiber, tote bags, furniture, carpet, paneling, straps, bottles and food containers (as long as the plastic being recycled meets purity standards and doesn›t have hazardous contaminants)

Symbol#2: HDPE

HDPE (high density polyethylene) is a versatile plastic with many uses, especially when it comes to packaging. It carries low risk of leaching and is readily recyclable into many types of goods.

Found in: Milk jugs; juice bottles; bleach, detergent, and other household cleaner bottles; shampoo bottles; some trash and shopping bags; motor oil bottles; butter and yogurt tubs; cereal box liners

How to recycle it: HDPE can often be picked up through most curbside recycling programs, although some allow only containers with necks. Flimsy plastics (like grocery bags and plastic wrap) usually can't be recycled, but some stores will collect and recycle them.

Recycled into: Laundry detergent bottles, oil bottles, pens, recycling containers, floor tile, drainage pipe, lumber, benches, doghouses, picnic tables, fencing, shampoo bottles

Symbol#3: PVC or V

PVC (polyvinyl chloride) and V (vinyl) is tough and weathers well, so it's commonly used for things like piping and siding. PVC is also cheap, so it's found in plenty of products and packaging. Because chlorine is part of PVC, it can result in the release of highly dangerous dioxins during manufacturing. Remember to never burn PVC, because it releases toxins.

Found in: Shampoo and cooking oil bottles, blister packaging, wire jacketing, siding, windows, piping

How to recycle it: PVC and V can rarely be recycled, but it's accepted by some plastic lumber makers. If you need to dispose of either material, ask your local waste management to see if you should put it in the trash or drop it off at a collection center.

Recycled into: Decks, paneling, mud-flaps, roadway gutters, flooring, cables, speed bumps, mats

Symbol#4: LDPE

LDPE (low density polyethylene) is a flexible plastic with many applications. Historically, it hasn't been accepted through most American recycling programs, but more and more communities are starting to accept it.

Found in: Squeezable bottles; bread, frozen food, dry cleaning, and shopping bags; tote bags; furniture

How to recycle it: LDPE is not often recycled through curbside programs, but some communities might accept it. That means anything made with LDPE (like toothpaste tubes) can be thrown in the trash. Just like we mentioned under HDPE, plastic shopping bags can often be returned to stores for recycling.

Recycled into: Trash can liners and cans, compost bins, shipping envelopes, paneling, lumber, landscaping ties, floor tile

Symbol#5: PP

PP (polypropylene) has a high melting point, so it's often chosen for containers that will hold hot liquid. It's gradually becoming more accepted by recyclers.

Found in: Some yogurt containers, syrup and medicine bottles, caps, straws

How to recycle it: PP can be recycled through some curbside programs, just don't forget to make sure there's no food left inside. It's best to throw loose caps into the garbage since they easily slip through screens during recycling and end up as trash anyways.

Recycled into: Signal lights, battery cables, brooms, brushes, auto battery cases, ice scrapers, landscape borders, bicycle racks, rakes, bins, pallets, trays

Symbol#6: PS

PS (polystyrene) can be made into rigid or foam products — in the latter case it is popularly known as the trademark Styrofoam. Styrene monomer (a type of molecule) can leach into foods and is a possible human carcinogen, while styrene oxide is classified as a probable carcinogen. The material was long on environmentalists' hit lists for dispersing widely across the landscape, and for being notoriously difficult to recycle. Most places still don't accept it in foam forms because it's 98% air.

Found in: Disposable plates and cups, meat trays, egg cartons, carry-out containers, aspirin bottles, compact disc cases

How to recycle it: Not many curbside recycling programs accept PS in the form of rigid plastics (and many manufacturers have switched to using PET instead). Since foam products tend to break apart into smaller pieces, you should place them in a bag, squeeze out the air, and tie it up before putting it in the trash to prevent pellets from dispersing.

Recycled into: Insulation, light switch plates, egg cartons, vents, rulers, foam packing, carry-out containers

Symbol#7: MISCELLANEOUS

A wide variety of plastic resins that don't fit into the previous categories are lumped into this one. Polycarbonate is number seven plastic, and it's the hard plastic that has worried parents after studies have shown it as a hormone disruptor. PLA (polylactic acid), which is made from plants and is carbon neutral, also falls into this category.

Found in: Three- and five-gallon water bottles, bullet-proof materials, sunglasses, DVDs, iPod and computer cases, signs and displays, certain food containers, nylon

How to recycle it: These other plastics are traditionally not recycled, so don't expect your local provider to accept them. The best option is to consult your municipality's website for specific instructions.

Recycled into: Plastic lumber and custom-made products

Compostable

Compostable
Products certified to be industrially compostable according to the European standard EN 13432/14955 may bear the 'seedling' logo.

Never place compostable plastic into the recycling with other plastics; as it is designed to break down it cannot be recycled and contaminates recyclable plastics. Plastics that carry this symbol can be recycled with your garden waste through your local authority.

Waste electricals
This symbol explains that you should not place the electrical item in the general waste. Electrical items can be recycled through a number of channels.

CHAPTER 8

Top Ten Award International Network Environmental Pioneers

Top Ten Award international Network (TTAIN) was established in 2012 to recognize outstanding individuals, groups, companies, organizations representing the best in the public works profession. TTAIN publishing books related to international Eco-labeling plans to increase public knowledge in purchasing based on the environmental impacts of products. We introduce in each volume some of the organizations that are doing their best in relation to taking care of the environmnet.

Argentina

LETIS is a company dedicated to the certification and auditing of international standards in quality and sustainability.

With an interdisciplinary team of professionals specialized in each industry and sector, LETIS works for local and global customers, with a vocation for innovation and personalized customer service.

LETIS team is committed to providing the highest quality service in a personalized way, generating long-term relationships with our clients and the community. Our certification system, our extensive network of producers, processors and traders, our presence in international trade shows and seminars, make LETIS a link to the world.

Contact information:
info@letis.org
Tel: 0341 5282560
www.letis.org

USA

1000 Springs Mill in Buhl, Idaho - locally owned, third generation Organic and NON-GMO organization producing products to benefit our local community. Our passion is Healthy People, Healthy Future. We offer a wide variety of products from retail ready to bulk packaging please visit website for further information. All our products are grown locally and manufactured at our facility. We also offer cleaning/milling services for Non-GMO & Organic commodities.

Contact:
1000 Springs Mill
430 7th Ave South
Buhl, ID 83316

Kurt Mason- 1000 Springs Mill Owner
E-mail- kurt@1000springsmill.com

Tim Cornie- 1000 Springs Mill Owner
E-mail- tim@1000springsmill.com

Paige Yore – Sales and Marketing
E-mail- paige@1000springsmill.com

Abra Snow- Organics Department
E-mail- abra@1000springsmill.com

CANADA

Our company summary: GreenCentury® was founded in 2008 by Harry Cheung, a Vancouver based company offering packaging products for the food industry. As a company, we pride ourselves on providing superior quality products and creating win-win long-lasting relationships, and treating all our clients with the same level of respect. A family run company by Harry and Maggie Cheung, who's previous business was the largest seller of DVD machines in Western Canada before that technology became obsolete.

Our Vision:
Doing business is adding value to each person we encounter with integrity and honour.

Our Mission:
We empower organizations toward full circle economy with sustainable products.

Our Values:
Sustainability, Community, Quality.

Contact:
Green Century Enterprises Inc.
Unit 113 - 7727 Beedie Way,
Delta, B.C. V4G 0A5. Canada
Tel: 604.214.7719
Fax: 604.303.7331
http://www.greencentury.ca

UNEP

The United Nations Environment Programme (UNEP) is the leading global environmental authority that sets the global environmental agenda, promotes the coherent implementation of the environmental dimension of sustainable development within the United Nations system, and serves as an authoritative advocate for the global environment.

Our mission is to provide leadership and encourage partnership in caring for the environment by inspiring, informing, and enabling nations and peoples to improve their quality of life without compromising that of future generations.

Headquartered in Nairobi, Kenya, we work through our divisions as well as our regional, liaison and out-posted offices and a growing network of collaborating centres of excellence. We also host several environmental conventions, secretariats and inter-agency coordinating bodies. UN Environment is led by our Executive Director.

We categorize our work into seven broad thematic areas: climate change, disasters and conflicts, ecosystem management, environmental governance, chemicals and waste, resource efficiency, and environment under review. In all of our work, we maintain our overarching commitment to sustainability.

Website: www.unep.org

RECYCLE SIGN
flat style

RECYCLE SIGN
line style

COMPOSTABLE
organic waste only

RECYCLE & REUSE
electrical goods

RECYCLABLE
aluminium ony

BIODEGRADABLE
selective waste

**PETE, HDPE, V, LDPE,
PP, PS, OTHER**

RECYCLE & REUSE
packaging materials

USE THE BIN
please don't litter

Bibliography

Bibliography:

Abramovitz, J.1998. Taking a stand: Cultivating a new relationship with the world's forests. Washington DC., Worldwatch 140: 84 p.

Agency (742-R-94-001 April).

Ahmad, M. 1998. Eco-Labeling of Indonesian Timber and Timber Products. Manila, Asian Development Bank, 13p.

Amount of Fraud. Journal of Law and Economics 16, 67-88.

Andrews, R.N.L. 1998. Environmental regulation and business 'self-regulation'. Policy Sciences 31(3): 177-197.

Apodaca, Julia, "Market Potential of Organically Grown Cotton as a Niche Crop." Natural Fibers Research and Information Center, Bureau of Business Research, University of Texas at Austin, Paper presented at the Beltwide Cotton Conference in Nashville, TN, January 1992.

Asadi, J., "International Environmental Labelling, Economic Consequencies, Export Magazine, July 2001

Asadi, J. 2008. Mobile Phone as management systems tools, ISO Magazine, Vol.8, No.1

Asadi, J., Eco-Labelling Standards, National Standard Magazine, Sep. 2004.

Assocs., Cambridge MA and G. Davis, U. Tenn, Knoxville,TN. (68-W6-0021): xiii+76+226pp.

Balter, M. 1999. Scientific cross-claims fly in continuing beef war. Science (May 28) 284: 1453-1455.

Belsley, D.A., Kuh, E., and Welsch, R.E. (1980), Regression Diagnostics, New York: John Wiley & Sons, Inc.

Birett, M. J. 1997. Encouraging Green Procurement Practices in Business: A Canadian Case Study in Program Development (108-118). in Greener Purchasing : Opportunities and Innovation. Sheffield, Greenleaf Publishing 325p.

Bowen, Nicola, World Agrochemical Markets, PJB Publications Ltd., March 1991.

Burnside, A., (1990), Keen on Green, Marketing, 17 May, pp35-36

Butler, D., (1990), A Deeper Shade of Green, Management Today, June, pp74-79

Cairncross, F. 1995. Green, Inc.: A guide to business and the environment. London, Earthscan. 277p.

Cason, T. N. and L. Gangadharan, (2002), Environmental Labeling and Charter, M. (ed.) 1992. Greener marketing: a responsible approach to business. Sheffield, Greenleaf Publishing 403p.

Chemical Week, 1999. Europe's Beef Ban Tests Precautionary Principle. (August 11).

CHOI, J.P. Brand Extension as Informational Leverage. Review of Eco- nomic Studies, Vol. 65 (1998), pp. 655-669.

Conway, G. 2000. Genetically modified crops: risks and promise.

Corrado, M., (1989), The Greening Consumer in Britain, MORI, London

Corrado, M., (1997), Green Behaviour – Sustainable Trends, Sustainable Lives?, MORI, london, accessed via countries. Manila, Asian Development Bank 33p.

Cropper, M.L., L.D. Deck, and K.E. McConnell. "On the choice of Functional Forms for Hedonic Price Functions," Review of Economics and Statistics 70(1988): 668-675.

Darbi, M. R. and E. Karni, (1973), Free Competition and the Optimal

Davis, G. 1998. Environmental Labeling Issues, Policies, and Practices Worldwide. Washington, DC. EPA, 216p.

Dawkins, K. 1996. Eco-labeling: consumer's right-to-know or restrictive business practice? Minneapolis, Minn., Institute for Agriculture and Trade Policy.

Di Leva, C. E. 1998. International Environmental Law and Development. Georgetown Interna. Environ. Law Review 10 (2): 502-549.

Economics and Management 43, 339-359.

Eiderstroem, E. 1997. Eco-labeling: Swedish Style. Forum for Applied Research in Public Policy 141(4).

Elkington, J. and Hailes, J. 1990. The green consumer guide: You can buy products that don't cost the earth. New York, Viking Press. 96p.

EMONS, W. Credence Goods and Fraudulent Experts. RAND Journal of Economics, Vol. 28 (1997), pp. 107-119.

EMONS, W. Credence Goods Monopolists. International Journal of In- dustrial Organization, Vol. 19 (2001), pp. 375-389.

Environment Canada 1997. Towards Greener Government Procurement: An Environment Canada Case Study (pp. 31-46). in Greener Purchasing: Opportunities and Innovations.

Environmental Protection Agency 742-R-98-009, (1998),

Environmentalist 17 (2): 125-133.

Erskine, C.C. and Collins, L. 1996. Eco-labeling in the EU: a comparative study of the pulp and paper industry in the UK and Sweden. European Environment 17 (2) : 40-47.

Erskine, C.C. and Collins, L. 1997. "Eco-labeling: Success or failure?".

Ethical Consumer, (1995), Co-op Supermarkets take up Ethics, EC36, June/July, p4

Ethical Consumer, (June 1996), Green Cons, EC41, June, p5

European Communities, Commission of the, 1996. Eco-label revision.

European Communities, Commission of the. 1996. Conservation of West Africa's forests through certification. UN Courier 157: 71-73.

FAO, 1999. State of the World's Forests 1999.

FAO, 1999. Wood Fuel Surveys.

Feenstra, R.C. "Exact Hedonic Price Indexes," Review of Economics and Statistics 77 (1995): 634-653.

Feenstra, R.C., and J.A. Levinsohn. "Estimating Markups and Market Conduct with Multidimensional Product Attributes," Review of Economic Studies (62 (1995): 19-52.

Forest Stewardship Council: "Principles and criteria for forest stewardship" Document 1.2: <http://www.fscoax.org>

Forsyth, K. 1999. Will consumers pay more for certified wood products? Journal of Forestry 97 (2) : 18-22.

Freeman, A. M III. The Measurement of Environmental and Resource Values. Theory and Methods. Washington D.C.: Resource for the Future, 1993.

Friends of the Earth, 1993. Timber certification and eco-labeling. London, FOE:

Graves, P., J.C. Murdoch, M.A. Thayer, and D. Waldman. "The Robustness of Hedonic Price Estimation: Urban Air Quality," Land Economics 64(1988): 220-233.

Halvorsen, R. and R. Palmquist. "The Interpretation of Dummy Variables in Semilogarithmic Equations." American Economic Review 70:474-75 (1980).

Imhoff, Dan, and Grose, Lynda, and Carra, Roberto., "Organic Cotton Exhibit," Mimeo. Simple Life and distributed the Texas Organic Cotton Marketing Cooperative, O'Donnell, Texas (1996).

Imhoff, Dan. "Growing Pains: Organic Cotton Tests the Fiber of Growers and Manufacturers Alike," reprinted on Simple Life's web page (simplelife.com), but first printed by Farmer to Farmer, December 1995.

Incomplete Consumer Information in Laboratory Markets. Journal of Environmental labeling.

ISO 14020, ISO 14021,ISO 14024,ISO 14025, International Organization for Standardization.

Kennedy, P.E. "Estimation with Correctly Interpreted Dummy Variables in Semilogarithmic Equations," American Economic Review 71: 801 (1981).

Kirchho®, S., (2000), Green Business and Blue Angels.

Kraus, Jeff. Lab Technician at the North Carolina School of Textiles.

Labeling Issues, Policies and Practices Worldwide.

Lamport, L. 1998. The cast of (timber) certifiers: who are they? International J. Ecoforestry 11(4): 118-122.

Large Scale impoverishment of Amazonian forests by logging and fire. 1999.

Lathrop, K.W. and Centner, T.J. 1998. Eco-labeling and ISO 14000: An analysis of US regulatory systems and issues concerning adoption of type II standards. Environmental

Lee, J. et al. 1996. Trade related environmental measures; sizing and comparing impacts.

Lehtonen, Markku. 1997. Criteria in Environmental Labeling: A comparative Analysis on Environmental Criteria in Selected Labeling Schemes. Geneva, UNEP. 148p.

LIEBI, T. Trusting Labels: A Matter of Numbers? Working Paper Uni versity of Bern, No. 0201 (2002).

Lindstrom, T. 1999. Forest Certification: The View from Europe's NIPFs. Journal of Forestry 97(3): 25-31. London

Losey, J.E., Rayor, L.S. & Carter, M.E. 1999. Transgenic pollen harms monarch larvae. Nature 399 20 May): p.214.

Management 22 (2) : 163-172.

Mattoo, A. and H. V. Singh, (1994), Eco-Labelling: Policy Considera-

Michaels, R. G., and V. K. Smith. "Market Segmentation And Valuing Amenities With Hedonic Models: The Case Of Hazardous Waste Sites," Journal of Urban Economics, 1990 28(2), 223-242.

Mintel, (1991), The Green Consumer I, May

Mintel, (1994), The Green Consumer, Mintel Special Report

Moraga-Gonzalez, J. L. and N. Padr¶on-Fumero, (2002),

NCC, (1996a), Green Claims – a consumer investigation into marketing claims about the environment,

NCC, (1996b), Shades of Green – consumers' attitudes to green shopping, National Consumer Council,

Nelson , P."Information and Consumer Behaviour," Journal of Political Economy 78 (1970): 311-329..

Nicholson-Lord, D., (1993) 'Tis the Season to be Green, The Independent, 20 December

Nuttall, N., (1993), Shoppers can cross green products off their lists, The Times, 3 July

OCDE/GD(97)105. Paris, OECD. 81p.

OECD. "Ec-labelling: Actual Effects of Selected Programmes," OCDE/GD (97) 105, 1997, Paris. (available on line at http://www.oecd.org/env/eco/books.htm#trademono)

OECD. 1997a. Case study on eco-labeling schemes. Paris, OECD (30 Dec):

OECD. 1997b. Eco-labeling: Actual Effects of Selected Programs.

Osborne, L. "Market Structure, Hedonic Models, and the Valuation of Environmental Amenities." Unpublished Ph.D. dissertation. North Carolina State University, 1995.

Osborne, L., and V. K. Smith. "Environmental Amenities, Product Differentiation, and market Power," Mimeo, 1997.

Ozanne, L.K. and Vlosky, R.P. 1996. Wood products environmental certification: the United States perspective". Forestry Chronicle 72 (2) : 157-165.

Palmquist, R. B., F. M. Roka, and T.Vukina. "Hog Operations, Environmental Effects, and Residential Property Values," Land Economics 73(1), (1997): 114-24.

Palmquist, R.B. "Hedonic Methods," in J.B Braden and C.D. Kolstad, eds. Measuring the Demand for Environmental Improvement. Amsterdam, NL: Elsevier, 1991.

Pento, T. 1997. Implementation of Public Green Procurement Programs (22-31) in Greener Purchasing: Opportunities and Innovations. Sheffield, Greenleaf Publ. 325 p.

Perloff, J. "Industrial Organization Lecture Notes," Mimeo. University of California at Berkeley (1985).

Plant, C. and Plant, J. 1991. Green business: hope or hoax? Philadelphia, New Society Publishers 136 p.

Polak, J. and Bergholm, K. 1997. Eco-labeling and trade: a cooperative approach (Jan.): Policy in a Green Market. Environmental and Resource Economics 22, 419-

Poore, M.E.D. et al. 1989. No timber without trees. London, Earthscan. 352p.

Raff, D. M.G., and M. Trajtenberg. "Quality-Adjusted Prices for the American Automobile Industry: 1906-1940." NBER Working Paper Series, Working Paper No. 5035, February 1995.

Rastogi, J. 1998. What's Behind the Label? Complexities of Certified Wood. Ecoforestry 13 (2): 38-42.

Roberts, J. T. 1998. Emerging global environment standards: prospects and perils. Journal of Developing Societies 14 (1): 144-163.

Rosen, S., "Hedonic Prices and Implicit Markets: Product Differentiation in Pure Competition." Journal of Political Economy. 82: 34-55 (1974).

Ross, B. 1997. Eco-friendly procurement training course for UN HCR. : 126 p.

Ryan, S., and Skipworth, M., (1993), Consumers turn their backs on green revolution, The Times, 4 April

Salzman, J. 1997. Informing the Green Consumer: The Debate over the Use and Abuse of Environmental Labels. Journal of Industrial Ecology 1 (2): 11-22.

Sanders, W. 1997. Environmentally Preferable Purchasing: The US Experience (946-960) in Greener Purchasing: Opportunities and Innovations. Sheffield, Greenleaf Publ. 325p.

Sayre, D. 1996. Inside ISO 14000: The competitive advantage of environmental management. Delray Beach FL., St. Lucie Press. 232p.

SHAPIRO, C. Premiums for High Quality Products as Returns to Reputa- tion. Quarterly Journal of Economics, Vol. 98, No. 4 (1983), pp. 659-680.

Stillwell, M. and van Dyke, B. 1999. An activists handbook on genetically modified organisms and the WTO. Washington DC., The Consumer's Choice Council: 20 p.

Teisl, M. F., B. Roe, and R. L. Hicks. "Can Eco-labels tune a market? Evidence from dolphin-safe labeling," Presented paper at the 1997 American Agricultural Economics Association Meetings, Toronto.

THE GERSEN, C. Psychological Determinants of Paying Attention to Eco- Labels in Purchase Decisions: Model Development and Multinational Vali- dation. Journal of Consumer Policy, Vol. 23, No. 4 (2000), pp. 285-313.

Tibor, T. and Feldman, I. 1995. ISO 14000: a guide to the new environmental management standards. Burr Ridge Ill., Irwin Professional Publ. 250 p.

Torre, I. de la, & Batker, D. K. (n.d.) 1999-2000. Prawn to trade: prawn to consume. Graham WA., Industrial Shrimp Action Network (isatorre@seanet.com), [and] Asia –Pacific

Townsend, M. 1998. Making things greener: motivations and influences in the greening of manufacturing. Aldershot, England, Ashgate Publisher. 203p.

U.S. Environmental Protection Agency. National Water Quality Fact Inventory: 1990 Report to Congress. EPA 503-9-92-006, Apr. 1992.

UK Eco-labelling Board website, accessed via http://www.ecosite.co.uk/Ecolabel-UK/

US Environmental Protection Agency (EPA742-R-99-001): 40 p. <www.epa.gov/opptintr/epp>

US EPA, 1993. Determinants of effectiveness for environmental certification and labeling programs. Washington, D.C., US Environmental Protect

US EPA, 1993. Status report on the use of environmental labels worldwide. Washington, D.C., US Environmental Protection Agency (742-R-93-001 September).

US EPA, 1993. The use of life-cycle assessment in environmental labeling. Washington, D.C., US Environmental Protection Agency (742-R-93-003 September).

US EPA, 1998. Environmental labeling: issues, policies, and practices worldwide. Washington DC., Environmental Protection Agency, Pollution Prevention Division Prepared by Abt

US EPA, 1999. Comprehensive procurement guidelines (CPG) program. Washington, D.C., US Environmental Protection Agency: <www.epa.gov/cpg>

US EPA, 1999. Environmentally preferable purchasing program: Private sector pioneers: How companies are incorporating environmentally preferable purchases. Washington, D.C.,

USG, 1993. Federal acquisition, recycling, and waste prevention. Washington DC., Executive Order: (20 October).

USG, 1998. Greening the government through waste prevention, recycling, and federal acquisition. Washington, D.C., Executive Order 13101 (September).

Van der Grijp, N. 1998. The Greening of Public Procurement in the Netherlands (60-71) in Greener Purchasing: Opportunities and Innovations. Sheffield, Greenleaf Pub. 325 p.

Vanclay, J.K. 1996. Lessons from the Queensland rainforests: steps towards sustainability. J. Sustainable Forestry 3 (2/3): 1-25.

Vidal, J., (1993), Shopping for a paler shade of green, The Guardian, 7 April

Voluntary Overcompliance. Journal of Economic Behavior and Organization

Von Felbert, D. 1995. Trade, environment and aid. Paris, OECD Observer 195: 6-10.

Ward, H. 1997. Review of European Community and International Environmental Law 6 (2): 139-147.

Wasik, John, F. Green Marketing and Management: a Global Perspective, Blackwell Business: Cambridge, Mass, 1996.

West, K. 1995. Ecolabels: the industrialization of environmental standards. The Ecologist (Jan/Feb) 25: 16-20.

Worcester, R., (1995), Business and the Environment – in the aftermath of Brent Spar and BSE, MORI,

World Commission on Forests and Sustainable Development: Final Report. <http://iisd.ca/wcfsd>.

Zarrilli, S., V. Jha, and R. Vossenaar, eds. Eco-labelling and International Trade, St martin Press, Inc. New-York, 1997.

Appendix I

Search by logos

Here you can search the logos in this volume. It will help you to better undersand the Ecolabels you may encounter while shopping. Buying Eco-products will aid in having a better environment with minimum polution during production processes. Three important parameteres for shopping are **quality**, **price** & **environmental impacts** of the products.

Goto page: 77	Goto page: 48

Goto page: 41	Goto page: 39

Goto page: 34	Goto page: 29

Goto page: 40	Goto page: 75

INTERNATIONAL ENVIRONMENTAL LABELLING VOL.1 • 95

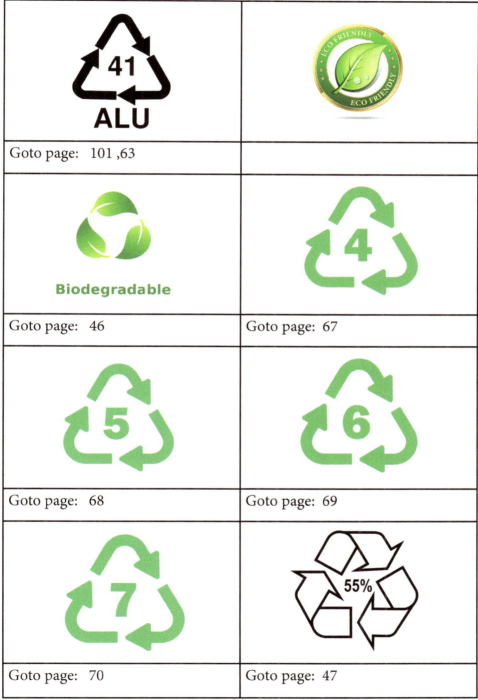

Goto page: 101, 63	
Goto page: 46	Goto page: 67
Goto page: 68	Goto page: 69
Goto page: 70	Goto page: 47

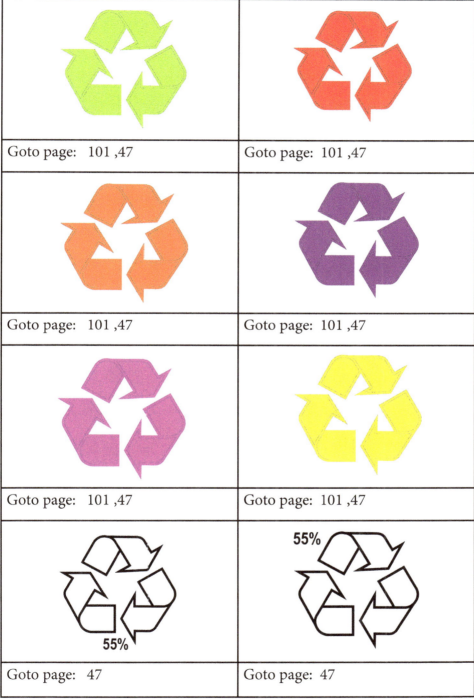

Appendix II

Recycling Codes

Recycling codes are used to identify the material from which an item is made, to facilitate easier recycling or other reprocessing. The presence on an item of a recycling code, a chasing arrows logo, or a resin code, is not an automatic indicator that a material is recyclable; it is an explanation of what the item is made of. Codes have been developed for batteries, biomatter/organic material, glass, metals, paper, and plastics.[citation needed] Various countries have adopted different codes. For example, the table below shows the polymer resin (plastic) codes. In the United States there are fewer, because ABS is placed with "others" in group 7.

A number of countries have a more granular system of recycling codes. For example, China's polymer identification system has seven different classifications of plastic, five different symbols for post-consumer paths, and 140 identification codes. The lack of a code system in some countries has encouraged those who fabricate their own plastic products, such as RepRap and other prosumer 3-D printer users, to adopt a voluntary recycling code based on the more comprehensive Chinese system.

RECYCLING CODES

PLASTIC: 01 PET, 02 PE-HD, 03 PVC, 04 PE-LD, 05 PP, 06 PS, 07 O, PA, ABS

BATTERIES: 08, 09, 10, 11, 12, 13, 14

PAPER: 20 PAP, 21 PAP, 22 PAP

METALS: 40 FE, 41 ALU, alu

BIOMATTER: 50 FOR, 51 FOR, 60 COT, 61 TEX, 62 TEX, 63 TEX, 64 TEX, 65 TEX, 66 TEX

GLASS: 70 GL, 71 GL, 72 GL, 73 GL, 74 GL, 75 GL, 76 GL, 77 GL, 78 GL, 79 GL

BIOMATTER: 81 PAP PET, 82, 83, 84 C/PAP, 85, 87 CSL, 90 C/LDPE, 91 C/LDPE, 92, 95, 96, 97, 98

ORGANIC — PAPER — PLASTIC — GLASS — METAL — E-WASTE — MIXED

PLEASE RECYCLE

Appendix III

Environmental Friendly Photos

Environmental friendly photos will be placed in this appendix. These photos can be received in the Top Ten Award International Network inbox from anywhere and everywhere, all over the globe. You can send your appropriate photos to us for them to be considered for publishing in one of the future, related volumes. They will be published with proper credit to the sender. The pictures can also be images of the Ecolabels existing in products within your country.

International Environmental Labelling Next Volumes

Vol.2
Energy & Electrical

Vol.3
Fashion & Textile

Vol.4
Health & Beauty

Vol.5
Maintenance & Cleaning

Vol.6
Wood & Stationery

Vol.7
DIY & Construction

Vol.8
Agriculture & Gardening

Vol.9
Professional Products

Vol.10
Financial Services

Vol.11
Tourism & Leisure

CPSIA information can be obtained
at www.ICGtesting.com
Printed in the USA
LVHW011541301120
672965LV00035B/1273

Vol.1

For All Food Industries
(Meat, Beverage, Dairy, Bakeries, Tortilla, Grain and Oilseed, Fruit and Vegetable, Seafood, And Sugar and Confectionery)

INTERNATIONAL ENVIRONMENTAL LABELING

Vol.2

For All Energy & Electrical Industries
(Renewable Energy, Biofuels, Solar Power, Hydroelectric Heating & Cooling, Wind Power, Energy Conservation, Geothermal and Nuclear Power)

INTERNATIONAL ENVIRONMENTAL LABELING

Vol.3

For All Fashion & Textile Industries
(Fashion Design, The Fashion System, Fashion Retailing, Marketing and Merchandizing, Textile Design and Production, Clothing and Textile Recycling)

INTERNATIONAL ENVIRONMENTAL LABELING

Vol.4

For All Health & Beauty Industries
(Fragrances, Makeup, Cosmetics, Personal Care, Sunscreen, Toothpaste, Bathing, Nailcare & Shaving, Skin Care, Foot Care, Hair Care and Other Health & Beauty Products)

INTERNATIONAL ENVIRONMENTAL LABELING

Vol.5

For All Maintenance & Cleaning Products
(All-purpose Cleaners, Abrasive Cleaners, Powders, Liquids, Specialty Cleaners, Kitchen, Bathroom, Glass and Metal Cleaners, Bleaches, Disinfectants and Disinfectant Cleaners)

INTERNATIONAL ENVIRONMENTAL LABELING

Vol.6

For All Wood & Stationery Industries
(Wooden Products, Cardboard, Papers, Markers, Pens, Notebooks, Writing Pads and Writing Sets, Pencils, White Papers, Envelopes and Organizers, Staplers and Paper Clips)

INTERNATIONAL ENVIRONMENTAL LABELING

Vol.7

For All DIY & Construction Industries
(Do it yourself ("DIY") of Building, Modifying, or Repairing, Renovation, Construction Materials, Cement, Coarse Aggregates, Clay Bricks, Power Cables, Pipes and Fittings, Plywood, Tiles, Natural Flooring)

INTERNATIONAL ENVIRONMENTAL LABELING

Vol.8

For All Agriculture & Gardening Industries
(Shifting Cultivation, Nomadic Herding, Livestock Ranching, Commercial Plantations, Mixed Farming, Horticulture, Butterfly Gardens, Container Gardening, Demonstration Gardens, Organic Gardening)

INTERNATIONAL ENVIRONMENTAL LABELING

Top Ten Award International Network

Top Ten Award International Network (TTAIN) was established in 2012 to recognize outstanding individuals, groups, companies, organizations representing the best in the public works profession. TTAIN publishing books related to international Eco-labeling plans to increase public knowledge in purchasing based on the environmental impacts of products. Top Ten Award International Network provides A to Z book publishing services and distribution to over 39,000 booksellers worldwide, including Apple, Amazon, Barnes & Noble, Indigo, Google Play Books, and many more. Our services including: editing, design, distribution, marketing TTAIN Book publishing are in the following categories: Student, Standard, Business, Professional, Honorary. We focus on quality, environmental & food safety management systems , as well as environmental sustain for future kids. TTAIN also provide complete consulting services for QMS, EMS, FSMS, HACCP and Ecolabeling based on international standards. ISO 14024 establishes the principles and procedures for developing Type I environmental labelling programmes, including the selection of product categories, product environmental criteria and product function characteristics, and for assessing and demonstrating compliance. ISO 14024 also establishes the certification procedures for awarding the label. TTAIN has enough experiences to help create new ecolabeling programmes in different countries all over the world. For more detail contact us:

+1- (778) 358-9589

info@toptenaward.net

INTERNATIONAL ECO SHOPPING GUIDE

For All Supermarket Customers

Vol.11
For All Tourism Industries
(Airline Industry, Travel Agent, Car Rental, Water Transport, Coach Services, Railway, Spacecraft, Hotels, Shared Accommodation, Camping, Bed & Breakfast, Cruises, Tour Operators)

Vol.10
For All Financial Products & Services
(Banking, Professional Advisory, Wealth Management, Mutual Funds, Insurance, Stock Market, Treasury/Debt Instruments, Tax/Audit Consulting, Capital Restructuring, Portfolio Management)

Vol.9
For All Professional Products & Services
(Teachers, Pilots, Lawyers, Advertising Professionals, Architects, Accountants, Engineers, Consultants, Human Resources Specialist, R&D, Psychologists, Pharmacist, Commercial Banker, Research Analyst)